U.S. ENVIRONMENTAL PROTECTION AGENCY

OFFICE OF INSPECTOR GENERAL

Response to Congressional Inquiry Regarding the EPA's Emergency Order to the Range Resources Gas Drilling Company

Report No. 14-P-0044   December 20, 2013

Report Contributors:

Kathlene Butler
Dan Engelberg
Johnny Ross
Genevieve Borg Soule

Abbreviations

DOJ	U.S. Department of Justice
EPA	U.S. Environmental Protection Agency
ISE	Imminent and Substantial Endangerment
OECA	Office of Enforcement and Compliance Assurance
OIG	Office of Inspector General
RRC	Railroad Commission of Texas
SDWA	Safe Drinking Water Act

Cover Photo: Outside the Range Resources' Butler and Teal hydraulic fracturing well sites. (EPA OIG photo)

At a Glance

Why We Did This Review

We conducted this review in response to a congressional request. We evaluated the U.S. Environmental Protection Agency's (EPA's) Region 6 issuance and withdrawal of an emergency order under Section 1431 of the Safe Drinking Water Act to the Range Resources Gas Drilling Company, to determine whether the EPA followed applicable laws and policy.

Region 6 concluded that a gas well owned and operated by Range Resources in Parker County, Texas, either caused or contributed to contamination found in the groundwater. Subsequently, on December 7, 2010, Region 6 issued an emergency order instructing Range Resources to investigate the groundwater and soil in the contaminated area to determine the cause of the contamination and to take actions to remediate and prevent further contamination.

This report addresses the following EPA themes:

- *Making a visible difference in communities across the country.*
- *Taking action on toxics and chemical safety.*
- *Protecting water: A precious limited resource.*

For further information, contact our public affairs office at (202) 566-2391.

The full report is at:
www.epa.gov/oig/reports/2014/
20131220-14-P-0044.pdf

Response to Congressional Inquiry Regarding the EPA's Emergency Order to the Range Resources Gas Drilling Company

What We Found

Region 6's issuance of the emergency order to Range Resources under Section 1431 of the Safe Drinking Water Act, and the region's subsequent enforcement actions, conformed to agency guidelines, regulations and policy. The region's interactions with state officials and other stakeholders were appropriate and within Section 1431 guidelines.

The EPA withdrew its emergency order regarding Range Resources hydraulic fracturing operations, but continues to monitor the situation for evidence of widespread contamination.

Laws and guidance do not address withdrawing Section 1431 emergency orders and the EPA used its discretion in withdrawing the emergency order. The EPA reached an agreement whereby Range Resources agreed to test 20 water wells every 3 months for a year to provide information about the presence of more widespread contamination. According to the EPA, the sampling that Range Resources has completed indicates no widespread methane contamination of concern in the wells that were sampled in Parker County. However, the EPA lacks quality assurance information for the Range Resources' sampling program, and questions remain about the contamination.

Recommendations and Planned Corrective Actions

We recommend that the Region 6 Regional Administrator (1) collect and evaluate the testing results being provided by Range Resources to determine whether the data is of sufficient quality and utility, (2) determine whether an imminent and substantial endangerment still exists at the original residential well involved, (3) inform the affected residents of the present status of the contamination and of any Region 6 planned actions, (4) work with the Railroad Commission of Texas to ensure appropriate action is taken as needed, and (5) document the costs and resources invested to complete the work included in these recommendations.

In its official comments and in subsequent meetings, the EPA agreed with and provided corrective actions that address our recommendations. All recommendations are resolved with corrective actions underway. No final response to this report is required.

UNITED STATES ENVIRONMENTAL PROTECTION AGENCY
WASHINGTON, D.C. 20460

December 20, 2013

MEMORANDUM

SUBJECT: Response to Congressional Inquiry Regarding the EPA's Emergency Order to the
Range Resources Gas Drilling Company
Report No. 14-P-0044

FROM: Arthur A. Elkins Jr.

TO: Ron Curry, Regional Administrator
Region 6

Cynthia Giles, Assistant Administrator
Office of Enforcement and Compliance Assurance

This is our report on the U.S. Environmental Protection Agency's (EPA's) emergency order to the
Range Resources Gas Drilling Company conducted by the EPA Office of Inspector General (OIG). This
report contains findings that describe the problems the OIG has identified and corrective actions the OIG
recommends. This report represents the opinion of the OIG and does not necessarily represent the final
EPA position. Final determinations on matters in this report will be made by EPA managers in
accordance with established audit resolution procedures.

Action Required

All recommendations are agreed to and resolved. Therefore, no final response to this report is needed.
If you wish to provide a final response to this report, it should be provided as an Adobe PDF file that
complies with the accessibility requirements of Section 508 of the Rehabilitation Act of 1973, as
amended. The final response should not contain data that you do not want to be released to the public;
if your response contains such data, you should identify the data for redaction or removal along with
corresponding justification. We will post this report to our website at http://www.epa.gov/oig.

If you or your staff have any questions regarding this report, please contact Carolyn Copper,
Assistant Inspector General for Program Evaluation, at (202) 566-0829 or copper.carolyn@epa.gov; or
Dan Engelberg, Director, Water Program Evaluations, at (202) 566-0830 or engelberg.dan@epa.gov.

Table of Contents

Chapters

Appendices

Chapter 1
Introduction

Purpose

This assignment responds to a congressional request the Office of Inspector General (OIG) received from six United States Senators to evaluate the U.S. Environmental Protection Agency's (EPA's) issuance and withdrawal of an emergency order under the Safe Drinking Water Act (SDWA). Our objective was to determine whether EPA Region 6's issuance of an emergency order to the Range Resources Gas Drilling Company under Section 1431 of the SDWA, and the region's subsequent enforcement actions, conformed to agency guidelines, regulations and policy. We also reviewed the region's interactions with state officials and other stakeholders, and the EPA's withdrawal of the emergency order.

Background

SDWA Provides the EPA with Emergency Powers

Congress established the SDWA to protect the quality of drinking water in the United States. Although most of the SDWA is concerned with ensuring that drinking water meets standards at public drinking water systems, part of the law, including Section 1431, is directed toward protecting drinking water sources from contamination.

Section 1431 of the SDWA authorizes the EPA to take immediate action to protect public health when any source of drinking water is, or will be, contaminated when two conditions exist. First, the EPA has information that a contaminant is in or likely to enter a public water system or underground drinking water supply and may present an imminent and substantial endangerment (ISE) to public health. Second, state and local authorities have not acted to protect public health from the ISE.

The preventative nature of Section 1431 means that for the EPA to take and enforce a Section 1431 emergency order, it needs neither proof that contamination has already occurred nor proof that the recipient of the order is responsible for the contamination. EPA guidance says that the EPA may act when the ISE is either direct or indirect, and whether the ISE is foreseeable in the near future or present at the time.[1]

Case law has supported the EPA's authority under the emergency powers provided in Section 1431 to "overlook technological and economic feasibility...unlimited by other constraints, [to] giv[e] paramount importance to

---

[1] *Final Guidance on Emergency Authority under Section 1431 of the Safe Drinking Water Act.* September 27, 1991.

the sole objective of the public health."[2] Individuals at both the EPA and the U.S. Department of Justice (DOJ) explained that SDWA Section 1431 gives the EPA the authority to take action to address emergencies proactively, even when the EPA does not have comprehensive information about the scenario. As a result, court opinions and case law have tended to give the EPA deference in these cases. The EPA's guidance reads:

> Even though EPA should strive to create a record basis to support its Section 1431 actions, the Regions should recognize that EPA does not need uncontroverted proof that contaminants are present in or likely to enter the water supply or that an imminent and substantial endangerment may be present before taking action under Section 1431. Similarly, EPA does not need uncontroverted proof that the recipient of the order is the person responsible for the contamination or threatened contamination. Courts generally will give deference to EPA's technical findings of imminent and substantial endangerment.[3]

The EPA guidance says that if the responsible party is not clearly known, an emergency order should be issued to the most likely contributor(s) based on the type of contaminant(s) found as compared to current and past land practices in the area. When the EPA determines that the two Section 1431 conditions are met, the EPA may take the steps necessary to protect public health. As part of an order, the guidance states that the EPA can require that a study be performed to more clearly determine the responsible parties. The EPA may:

- Order those who caused or contributed to the endangerment provide alternative water supplies, at no cost to the consumers (e.g., provisions of bottled water, drilling of new well(s) and connecting to an existing public water system).
- Notify the public of hazards (e.g., door-to-door, posting, newspapers and electronic media).
- Order studies to determine the extent of the contamination.
- Order engineering studies to propose a remedy to the endangerment and a timetable for its implementation.
- Order the halting of the disposal of contaminants that may be contributing to the endangerment.

Section 1431 and the EPA guidance do not offer criteria for withdrawing an emergency order. Therefore, the EPA uses discretion to decide when to withdraw an emergency order.

---

[2] *United States v. Hooker Chem. & Plastics Corp.,*749 F.2d 968, 988 (2d Cir.1984).
[3] *Final Guidance on Emergency Authority under Section 1431 of the Safe Drinking Water Act.* September 27, 1991.

SDWA Advises Consulting With States on Emergency Orders

Most states have received authority to implement the SDWA requirements. However, the EPA retains responsibility to oversee drinking water programs by taking actions when states do not. Section 1431 of the SDWA instructs the EPA to consult with the state and local authorities prior to taking enforcement action "to the extent practicable."

The Texas Commission on Environmental Quality is the chief environmental agency for the state of Texas. However, for matters involving the production of oil and gas, the Railroad Commission of Texas (RRC) has regulatory authority. The RRC, through its Oil and Gas Division, regulates the exploration, production and transportation of oil and natural gas. Its statutory roles include preventing waste of the state's natural resources, protecting the correlative rights of different interest owners and preventing pollution. To prevent pollution of the state's surface and groundwater resources, the RRC has an abandoned well plugging and abandoned site remediation program. The oil and gas industry partially funds this program through fees and taxes. Within the Oil and Gas Division, the Site Remediation Section, with field offices throughout the state, accepts and investigates complaints of contamination caused by oil and gas production.

EPA Policy Requires Press Releases to List Statutes, Risks and Precedents

The EPA's *2007 EPA Policy on Publicizing Enforcement and Compliance Assurance Activities* provides guidance on the content of enforcement publications and the proper review process. The policy states that the EPA should not negotiate the content of press releases outside of the agency. It states that the press release should include the statute(s) violated, the environmental and health impacts of the specific pollutants or contaminants involved and whether the case would create national or program precedents. It also states that the relevant press office has ultimate responsibility for final editorial control over the content of national/regional press announcements and is responsible for disseminating them to the media, regions and program offices.

A review of the Information Quality Act and subsequent guidelines issued by both the Office of Management and Budget[4] and the EPA[5] indicates that press releases are exempt. The Information Quality Act does not apply to communications such as press releases.

---

[4] *Office of Management and Budget Guidelines for Ensuring and Maximizing the Quality, Objectivity, Utility, and Integrity of Information Disseminated by Federal Agencies; Republication.* February 22, 2002.
[5] *EPA Guidelines for Ensuring and Maximizing the Quality, Objectivity, Utility, and Integrity of Information Disseminated by the Environmental Protection Agency.* October 2002.

The EPA Investigated a Drinking Water Contamination Complaint and Issued an Emergency Order

In August 2010, a homeowner in Parker County, Texas, complained to Region 6 that the drinking water well associated with his home had become contaminated with natural gas and requested assistance. In a phone call to Region 6, the homeowner stated that the well pump malfunctioned because high levels of natural gas in the water caused the pump to lose suction. He reported that his drinking water was effervescing inside the home, indicating high levels of gas in the water. The homeowner indicated that he could set his drinking water on fire to illustrate high levels of natural gas in the water at the wellhead. He indicated that he had contacted state officials at the RRC, the state arm responsible for investigating contamination of drinking water wells, but they had not been able to resolve his issues.

In response to the complaint, on October 26, 2010, Region 6, in consultation with the RRC officials, conducted sampling and testing of the air and the well water at two residential wells to verify the existence and nature of the contamination. Region 6 also identified a nearby gas production well as a potential source and collected gas samples for isotopic and compositional analysis from both the gas well, operated by Range Resources, and the drinking water wells. The EPA conducted the analysis to determine the possible origin of the gases in the drinking water and gas wells, and to compare the composition of gas in the well water to gas from the production well.

Region 6 received results from its October sampling and testing on November 16, 2010. The testing results prompted the EPA to advise the residents at both homes to discontinue use of the well water. The test results showed levels of methane above action levels set by USGS (i.e. 10 mg/L, the level at which wells should be evaluated for venting and ignition sources should be removed from the area). This presented a potential explosion hazard. The test results also showed benzene levels above the EPA published maximum contamination levels. Based on an isotopic analysis, Region 6 concluded that gas in the groundwater and gas from the production well were nearly identical and likely originated from the same source. Region 6, therefore, concluded that a gas production well owned by Range Resources caused or contributed to the contamination in the groundwater. Region 6 provided its test results to officials at the RRC, telling the state that the evidence demonstrated an ISE.

The RRC informed Region 6 that it did not share the EPA's conclusion that the Range Resources gas well caused the drinking water well contamination, but said the RRC would continue its own testing and research. Region 6 requested a meeting with Range Resources to discuss its results as it related to the gas production well. However, Range Resources declined to meet with the region and indicated that they would be working with state officials to investigate and resolve the issue.

Based on the evidence collected and its discussions with the RRC in November 2010, Region 6 began coordinating with the EPA's headquarters to take emergency action under the SDWA. Region 6 issued an emergency order to Range Resources on December 7, 2010, citing that the gas production well either caused or contributed to the contamination in two residential water wells.[6] The order required Range Resources to conduct research on the source and extent of contamination, provide drinking water to affected residents, and develop a plan to mitigate contamination in the aquifer.

Range Resources did not fully comply with the order, and legal actions between the company and EPA ensued. EPA withdrew the order in March 2012, reaching a nonbinding agreement with Range Resources for additional well testing.

Scope and Methodology

We conducted our evaluation in accordance with generally accepted government auditing standards. Those standards require that we plan and perform our work to obtain sufficient, appropriate evidence to provide a reasonable basis for our findings and conclusions based on our audit objectives. We believe that the evidence obtained provides a reasonable basis for our findings and conclusions based on our evaluation objectives. We conducted our evaluation from July 2012 to July 2013.

We sought to determine whether Region 6's issuance, implementation and withdrawal of the emergency order under Section 1431 of the SDWA met all requirements of the Act. We also sought to:

- Determine whether Region 6's interaction with state officials, EPA's headquarters and other stakeholders was appropriate and in accordance with Section 1431 guidelines.
- Determine the applicability of the Information Quality Act to Region 6's press release concerning its Section 1431 actions.
- Determine whether Region 6's Section 1431 actions for Range Resources were compatible to other Section 1431 actions for other violators.
- Evaluate eight items enumerated in the congressional request letter dated June 19, 2012.

To complete our work, we conducted interviews and obtained and reviewed documents and official records. We interviewed staff and officials in the EPA's Office of Enforcement and Compliance Assurance (OECA), EPA's Region 6, the DOJ, the RRC and Range Resources. We interviewed homeowners at one contaminated residential well and their attorney. We made site visits to the homes,

---

[6] Subsequent to the emergency order, the EPA determined that it was unlikely that Range Resources' gas well drilling and production activities had caused the contamination in Residential Well 2.

to the contaminated water wells and to the two Range Resources gas production wells. We attempted to interview residents at two other homes where contaminated wells were identified, but they did not agree to participate in our evaluation. We reviewed the order, administrative record and court documents, and assessed evidence surrounding the contamination from the EPA, Range Resources and the RRC.

We compared the Range Resources emergency order with other emergency orders the EPA issued between January 1, 2009, and December 31, 2011. We used the EPA's Enforcement and Compliance History Online EPA enforcement cases search tool to query all SDWA Section 1431 orders issued during the 3-year period. This search returned 40 orders, the overwhelming majority of which were against public water systems and not relevant for comparison. Of the 40 orders, we identified two cases for comparison: one involved a company that was also part of the oil and gas industry, and the other involved a single company allegedly contaminating one or more private citizen water wells. We compared the elements of the Range Resources order with these two orders.

Chapter 2
The EPA Adhered to Policy, but Questions Remain

The EPA's issuance of the emergency order to Range Resources under SDWA Section 1431 and Region 6's subsequent enforcement actions conformed to agency guidelines, regulations and policy. The Range Resources order was similar to two other emergency orders the EPA issued in the past. The region's interactions with the RRC and other stakeholders followed Section 1431 guidelines. Laws and guidance do not address withdrawing Section 1431 emergency orders, so the EPA withdrew the order based on its discretion. Upon withdrawal, the EPA reached a non-binding agreement with Range Resources for additional testing in the area. According to the EPA, the sampling that Range Resources completed in wells in Parker County, Texas, indicated no widespread methane contamination at action levels[7] (i.e., one well of 20 showed methane above action levels, and a subsequent sample at this well was below the limit). However, the EPA lacks quality assurance information for Range Resources' sampling program, and questions remain regarding the presence of contamination.

Emergency Order Met the Two Requirements of SDWA

Section 1431 requires two conditions for the EPA to take emergency enforcement action. Both existed at the time Region 6 issued the emergency order. First, the EPA concluded that two residential drinking water wells were contaminated with methane, benzene and other contaminants. Second, test results showed methane levels that could accumulate in the affected homes and potentially cause an explosion.

The Contamination Warranted an Emergency Order

The EPA was justified in concluding that the contamination in the residential wells constituted an ISE based on the data the EPA collected. The EPA water and gas samples collected from the residential water wells were contaminated. Test results on November 16, 2010, showed the presence of chemical contamination in both wells. The contamination levels indicated a risk to a drinking water source—the aquifer and the wells drawing from it.

The U.S. Department of the Interior advises owners of wells with methane concentrations greater than 10 milligrams per liter to consider removing ignition sources from the immediate area to prevent the possibility of an explosion. The EPA used this standard to conclude that the concentration of methane in

---

[7] i.e., 10 milligrams per liter, the level at which wells should be evaluated for venting and ignition sources should be removed from the area.

residential water wells presented a risk of explosion. Data showed dissolved methane at one residential water well reaching twice this limit: 20.1 milligrams per liter. Data also showed benzene in one residential well at 6.84 micrograms per liter. This level is above the maximum SWDA contamination standard set for public water supplies of 5 micrograms per liter. In addition, results indicated the presence of ethane, propane, toluene and hexane.

The methane in the wells presented an explosion hazard, and benzene presented health hazards. Methane poses risks of explosion and fire. In large concentrations in air, it may pose a risk of asphyxiation. Benzene is a known human carcinogen. It can cause anemia, neurological impairment and other adverse health impacts. Hexane, propane, ethane and toluene may also cause adverse health impacts if inhaled or ingested. The residences housed nine people, including adults and young children. Region 6 staff concluded that the levels of methane and benzene in the water posed an imminent and substantial endangerment to the residents.

State and Local Authorities Did Not Plan to Act Immediately

Evidence shows that EPA correctly determined that the RRC did not plan to act immediately. Region 6 officials and staff communicated with the RRC officials and staff while they were investigating the contamination, and while the EPA was considering taking action. The EPA asked the RRC if they planned to take action, and the RRC said they were not prepared to do so. State laws did not authorize officials at the Texas Commission on Environmental Quality to act because the contamination involved gas and oil production, which is outside of their jurisdiction. SDWA Section 1431 and the EPA's guidance[8] do not define how to determine whether actions will happen in a timely fashion. Therefore, based on their assessment of the explosion risk and the RRC response to the EPA's questions on taking action, the EPA concluded that appropriate state and local officials had not taken sufficient action to address the endangerment and did not intend to take timely action.

The EPA Concluded That a Gas Well Was the Most Likely Contributor to the Contamination

The information that the EPA had in its possession was sufficient for it to conclude that the gas production well was the most likely contributor to the contamination of the aquifer that led to the ISE. It was the closest potential source of contamination to the contaminated drinking water well and Range Resources drilled the well shortly before the homeowners first reported the contamination. Moreover, the EPA data showed the composition of the gas contaminating the water wells to be nearly identical to that of the gas in the production well, and that the gas from both types of wells was likely from the same source.

---

[8] *Final Guidance on Emergency Authority under Section 1431 of the Safe Drinking Water Act.* September 27, 1991.

The EPA met the legal burden required for Section 1431 when it issued the emergency order against Range Resources. Section 1431 does not expect the EPA to delay acting to address an ISE until it is certain of the source of the endangerment. Rather, the EPA guidance states:

> In cases where the responsible party is not clearly known, the order should be issued to the most likely contributor(s) based on the type of contaminant(s) found in … [the] USDW [underground source of drinking water] compared to current and past land practices in the area.[9]

The Location and Timing of the Gas Well Operation Coincided with Well Contamination

The location and timing of Range Resources gas well drilling and operations contributed to the EPA's determination that a nearby gas production well caused or contributed to well contamination. The EPA identified two gas production wells close to the two contaminated water wells, both operated by Range Resources. These wells went into production in August 2009, and were the only two natural gas wells within 2,000 feet of the contaminated residential water wells. The gas production wells both included a vertical bore and a horizontal fracture bore. The residential wells were both within 500 feet in horizontal distance from the fractured track of the horizontal section of the gas production well bore and approximately 2,300 feet from the vertical bore.

Figure 1. Approximate Location of Gas Production Wells and Contaminated Residential Wells

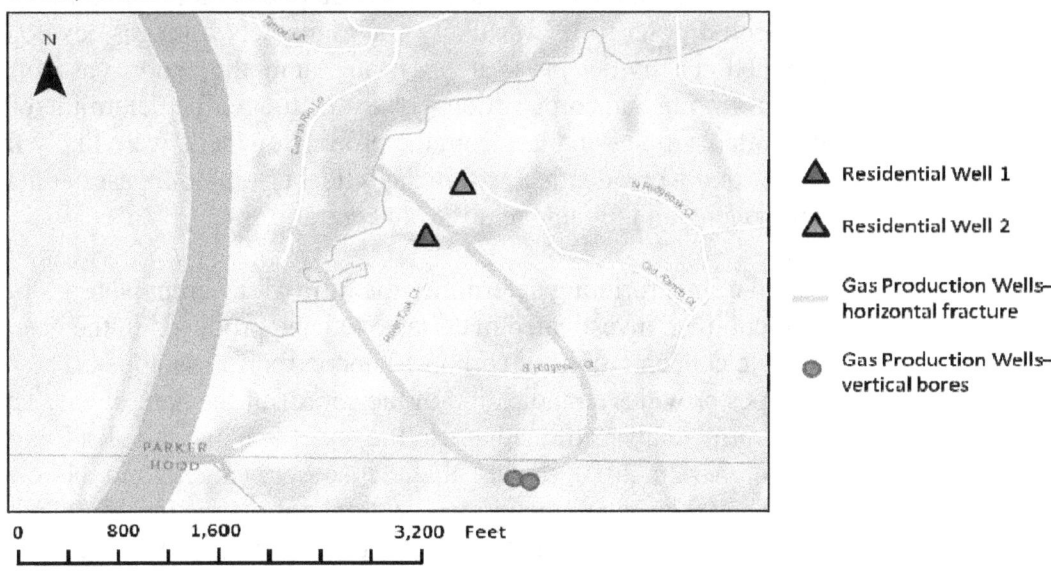

Source: EPA Region 6.

---

[9] *Final Guidance on Emergency Authority under Section 1431 of the Safe Drinking Water Act.* September 27, 1991.

The EPA relied on the coincident timing of the gas production well drilling and operations and the detection of the water well contamination to provide additional evidence of the connection. A professional drilling service drilled Residential Well 2 in August 2002 and Residential Well 1 in April 2005. The homeowners primarily used the water well for human consumption and landscape irrigation. Neither the homeowners nor the well drilling services observed or reported that either residential well contained any noticeable natural gas at the time of drilling. The water wells produced drinking water for years without signs of natural gas. In December 2009, approximately 4 months after Range Resources placed the gas wells into production, the homeowners at Residential Well 1 began noticing signs of natural gas. In May 2010, the homeowners at Residential Well 2 began noticing signs of natural gas.

Records indicate that Range Resources began drilling the gas production well in June 2009. The hydraulic fracture operation was completed and the well began gas production in August 2009. The homeowners reported noticing that drinking water started to effervesce (give off bubbles as gas escaped) in late 2009 and early 2010. In July 2010, the homeowners at Residential Well 1 reported that methane caused their pump to malfunction. At that point, the homeowners determined that the gas in the water well was flammable and contacted state officials and the EPA. This sequence of events contributed to Region 6's conclusion that the gas production well contributed to the contamination of the water well.

Isotopic Analysis Indicates Gases Nearly Identical and Likely From the Same Source

Based on data it collected, Region 6 concluded that the presence of gas in the residential water wells was likely due to impacts from gas development and production activities in the area. Through a method known as isotopic fingerprinting and compositional analysis, the region determined that the gas in the water samples and gas from the production wells were likely from the same source and were identical within analytical error. Both gases contained the same components in the same relative concentrations.

Isotopic fingerprinting determines the ratio of different isotopes of a particular element in an investigated material. Methane produced in the ground varies in the relative concentrations of carbon 12 and carbon 13 isotopes. The ratio of these isotopes provides an indication of the source of the natural gas. The isotopic fingerprint analysis of methane in the water and gas samples Region 6 obtained on October 26, 2010, showed that the isotopic values from the residential water wells and the gas production well were thermogenic in nature and likely from the same source. In addition, the compositional analysis indicated that both gases contain significant amounts of heavier hydrocarbon components, and that the gases were identical within analytical error; the hydrocarbon portion of each gas contains the same components.

Region 6 concluded that gases from the production wells and contaminated residential wells were likely from the same source. Moreover, samples taken from other wells in the area had a different isotopic profile. Region 6 concluded that the gas produced by Range Resources' gas well was most likely the source that was contaminating the aquifer and the drinking water wells. Figure 2 shows plotted isotopic values of gases from the gas production well and the residential water well along with gas samples from other nearby wells.

Figure 2. Isotopic Fingerprinting of Gas from Contaminated Residential and Gas Production Wells

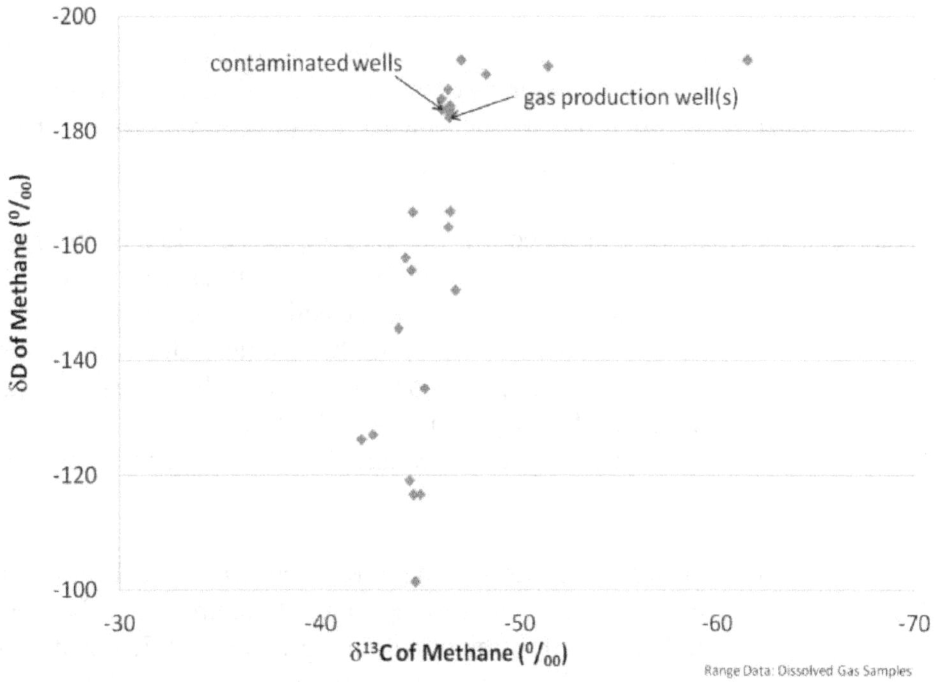

Source: EPA Region 6.

EPA Evidence Led to Emergency Order Against Range Resources

Based on the evidence the EPA uncovered regarding the nature and source of the contamination in the residential wells, the EPA determined that a Range Resources gas production well was the most likely contributor to the contamination. The EPA drew on its authority in SDWA Section 1431 to issue an emergency order to Range Resources to protect an underground source of drinking water from contamination. The EPA's emergency order outlined steps to investigate and remediate the contamination identified. Specifically, the emergency order directed the company to take the following steps:

- Provide drinking water for the consumers of the contaminated wells.
- Conduct research into potential contamination of drinking water supplies within 3,000 feet of the gas well.

- Conduct soil gas and indoor air concentration analysis at the houses served by the contaminated well.
- Identify and repair the contamination pathways.

OECA staff told us that after the order was issued, the EPA determined that it was unlikely that Range Resources' gas well drilling and production activities had caused the contamination in Residential Well 2. Subsequent discussions about contamination focused only on Residential Well 1.

Range Resources Emergency Order Similar to Other Emergency Orders

Our review of recent emergency orders revealed two recent situations similar to that of the Range Resources case. Both were situations where the EPA issued an emergency order in response to what it believed to be contamination threatening drinking water sources by a company.

- The first instance was the East Poplar Oil Field Section 1431 Order. This situation involved multiple companies drilling for oil in Montana. Region 8 determined that these activities had contributed to contamination of the groundwater. Although the public water system was not yet contaminated, Region 8 determined that there was an ISE. This is unlike the Range Resources case in that Region 6 believed that the private well was already contaminated and that there was an imminent potential of the well exploding.[10]

- The second instance was the Kenneth Brockett Farm Section 1431 Order. This case was similar to the Range Resources case because it involved a single company contaminating a private well. It was based on evidence that a dairy farm was contaminating a private well with fecal coliform, E. coli and ammonia at levels that presented an ISE.[11]

The EPA imposed similar requirements in all three cases. It ordered the companies to monitor the contamination in public water supplies and private wells, and to develop a mitigation or remediation plan to correct the situation. All three orders also directed companies to provide an alternate source of water for contaminated wells.

---

[10] East Poplar Oil Field, Docket No. SDWA-08-2011-00126, 12/16/10.
http://yosemite.epa.gov/oa/rhc/epaadmin.nsf/Filings/5BDBB00D74503113852578010020F7FA/$File/SDWA08201 10006%20AO.pdf.
[11] Kenneth Brockett Farm, Docket No. SDWA-03-2011-0205-EO, 6/28/11.
http://www.epa.gov/reg3wapd/pdf/public_notices/BrockettEmergencyOrder.pdf.

The Region's Communications With State and Other Stakeholders Adhered to Law, Policies and Procedures

Region 6's communications with the state satisfied the requirements of the SDWA its communications with the public and other relevant stakeholders adhered to agency policy and procedures. The region coordinated with its state partner, the RRC, as required under Section 1431. Region 6's press release regarding the Range Resources Section 1431 Emergency Order followed procedure in both content and distribution. The content of the release followed the EPA's policy and was similar to other press releases. The timing of the distribution followed agency procedures. Press releases are exempt from the Office of Management and Budget and the EPA's Information Quality Act guidelines. Therefore, these guidelines were not relevant for comparison.

The EPA Regularly Interacted With and Communicated With Its State Partners

The EPA followed Section 1431 by regularly consulting with the RRC prior to taking enforcement action. According to Region 6 documents, Region 6 staff consistently coordinated by phone and email with the RRC staff from August to December 2010 to discuss the investigation of the well water contamination and the issuance of the emergency order. Our review of records indicated that between August and October 2010 Region 6 staff had seven email exchanges with RRC staff located at the Abilene, Texas, field office. Between September and October 2010, Region 6 staff exchanged seven emails with the RRC Site Remediation office in Austin, Texas. This email traffic was concerning the contaminated well water.

The RRC staff told us that Region 6 staff coordinated with them via email and conference calls in an effort to collect samples from the gas and water wells and to discuss sampling methods. Subsequently, Region 6 and the RRC held a conference call to discuss the sampling method for the October 21, 2010, sampling. Region 6 and RRC staff visited the site and worked cooperatively in collecting gas and water samples. In November, Region 6 sent its test results of the October samples to the RRC staff. In addition, email traffic showed that Region 6 coordinated with the RRC to discuss the pending emergency order. Prior to issuing the order to Range Resources, Region 6 discussed the contents of the emergency order with the RRC.

Press Release Conformed to Requirements

The Region 6 press release regarding the Range Resources Section 1431 Emergency Order followed required procedure in both content and distribution. The region followed the typical procedures for distributing the press release to the public. The EPA issued the press release on December 7, 2010, the day it issued the emergency order. The region issued its press release 11 minutes prior to sending the order to Range Resources, which was not typical, but also not specifically against any policy provision. However, Region 6 has stated that Range Resources had knowledge of the order prior to its issuance on December 7, 2010. The EPA requested a meeting with Range Resources to discuss the impending order, which Range Resources declined. However, according to Region 6 enforcement personnel and Range Resources, the RRC provided Range Resources with details of the order prior to the EPA's issuing the order.

The contents of the release adhered to EPA policy and Region 6 followed EPA procedures in ensuring review and concurrence from headquarters. Additionally, the content was similar to other enforcement press releases we reviewed. The press releases for the two other Section 1431 cases that we reviewed also included additional information beyond the case facts. The EPA press personnel indicated that the agency does this to provide context for the enforcement action. They also indicated that this practice follows Associated Press standards. Finally, our review of the Information Quality Act and subsequent guidelines issued by both the Office of Management and Budget and the EPA found that press releases are exempt from the act; therefore, we did not assess EPA's Range Resources press release against these criteria.

The EPA Informed Citizens Groups About the Order

The former Region 6 Administrator informed environmental and citizen groups of the order and the associated press release after the region issued the two documents. Some members of these groups had shown interest in oil and gas issues in the state and had attended prior meetings with Region 6. A review of the evidence showed that this communication occurred after the region issued its press release and that it is not out of the ordinary for the EPA to inform interested parties of press releases after they are released. Although OECA's communication policy restricts discussions concerning press releases and administrative orders when they are in draft form, it places no such restrictions once they are issued.

Range Resources Did Not Fully Comply With the Order

Range Resources contended that the order was factually and legally unsupportable. In its response to the emergency order, Range Resources said that the company's investigation indicated that natural gas had been present in this aquifer long before Range Resources' production activities and was likely naturally occurring migration from several other shallow gas zones immediately below the aquifer. However, Range Resources responded to the December 2010 order by indicating that they had offered to provide water to the affected well owners. Range Resources did not comply with the other elements of the order.

Range Resources emphasized that the company drilled the gas production wells into the Barnett Shale formation, more than a mile below the aquifer from which the residential water well was drawing water. They said this demonstrated that the depth of the gas well's horizontal bore was vertically too far below the maximum water well depth to contaminate it with gas from the Barnett Shale. However, Region 6 countered by offering evidence from scientific literature suggesting that even at greater depths deep production wells may affect water wells in a shallow aquifer. Range Resources met with Region 6 staff on December 15, 2010, to discuss the emergency order, but Range Resources did not agree to comply with the EPA's order.

Subsequent Federal, State and Legal Actions Questioned Cause of Contamination

Several legal actions between the EPA and Range Resources regarding the contamination ensued and were ongoing between January 2011 and March 2012. On January 19, 2011, the RRC held a hearing to make its own determination about whether Range Resources caused the contamination. In contrast to the EPA guidance concerning emergency orders, the RRC needed definitive proof that a direct source was causing the contamination in order to take action. When we asked staff/officials of the RRC what sort of evidence RRC required to determine if a direct connection existed, they told us that they did not know. They said the RRC has never had a case where they found a direct connection between an oil or gas well and a drinking water well.

In this case, the Texas Railroad Commissioners found that the Range Resources gas wells were not causing or contributing to the contamination of any Parker County wells. Subsequent to this RRC finding, Range Resources ceased complying with the two parts of the EPA emergency order to which it had agreed: (1) it stopped providing alternate water to the one home where it had been doing so, and (2) removed the explosivity monitor it had installed in the home. The EPA disagreed with the finding reached by the state at the RRC hearing. In addition, the agency had collected additional information supporting its claim that Range Resources' gas well caused the contamination.

The EPA Filed Suit to Enforce the Order and Range Resources Filed Judicial Review

The DOJ filed an action (a complaint) on January 18, 2011, in the U.S. District Court for the Northern District of Texas to require Range Resources to comply with the order, among other things. This action did not receive a ruling because, on January 20, 2011, Range Resources filed an action seeking judicial review of the order in the U.S. Court of Appeals for the Fifth Circuit. The District Court issued a stay of the DOJ action to enforce the order pending the outcome of the Court of Appeals' ruling on Range Resources' action seeking judicial review.

Withdrawal of the Emergency Order Was Within the EPA's Discretion, but Questions Remain

The EPA's withdrawal of the emergency order did not violate any regulation or policy. Because SDWA Section 1431 and the EPA's guidance do not provide criteria or a process for withdrawing Section 1431 emergency orders, EPA officials exercised discretion in withdrawing it.

The EPA and Range Resources began discussions to resolve the situation outside of the courts and came to an agreement in March 2012. As such, also in March 2012, EPA withdrew the order, the DOJ withdrew its action seeking to enforce the order in District Court and Range Resources withdrew its action seeking judicial review of the order in the Court of Appeals before the Court of Appeals ruled.

Region 6 and OECA staff and officials cited several reasons for withdrawing the order. First, the EPA wanted to reduce the costs and legal risks associated with the ongoing court cases. In addition, an EPA official indicated that the EPA believed that the risk faced by the residents at the well where contamination had first been found was reduced because the residents had obtained water from a separate source and were no longer using the well. Finally, the EPA was able to obtain Range Resources' agreement to participate in a national agency study of the relationship between hydraulic fracturing and drinking water contamination. Range Resources also agreed to sample 20 water wells in Parker County every 3 months for a year if the EPA withdrew the order.

The EPA Perceived High Litigation Risk and Cost

Senior OECA and Region 6 officials indicated to us they were willing to settle the case because it was more complicated than they had anticipated. EPA leadership determined in 2012 that although they still believed the statute supported the EPA's actions, the case with Range Resources required more resources than anticipated because the judge had called for the review and consideration of additional evidence not ordinarily included in such cases.

An OECA official told us that enforcing the order in the courts is usually a simple process because Section 1431 does not require absolute proof of contamination or cause. A senior DOJ attorney who worked on the case said that he believed that the emergency order was sufficient and could have been enforced through the courts. The EPA's Assistant Administrator for OECA also told the OIG that she was "very confident" in the enforcement case. Both the DOJ and OECA officials said that the EPA had enough evidence and support to enforce the order. Further, they explained that Congress designed the existing statutes to protect people; these statutes give the EPA the authority to take action to address emergencies, so court opinions and case law have tended to give the EPA deference.

However, the EPA believed its prospects in this case were uncertain. OECA officials told us that although they believed they were on firm ground there was always a risk that the judge could rule against the EPA. If that happened, it would risk establishing case law that could weaken the EPA's ability to enforce Section 1431 emergency orders in the future.

Because the court was requiring additional evidence beyond that provided in the EPA's administrative record, EPA officials said they would have to gather additional evidence and expert witness testimony at additional cost. As a result, OECA senior officials said that was not an efficient use of agency resources.

The EPA Believed the Risk to Homeowners Had Been Reduced

In interviews, an OECA official indicated to us that a factor that influenced the withdrawal decision was the belief that the risk faced by users of the remaining residential well covered by the emergency order had lessened since they issued the emergency order in 2010. The homeowner who owned the well of primary concern had begun purchasing water from an alternate source and was no longer using the contaminated well. Moreover, in responding to our preliminary findings, OECA officials explained they also believed that the risk had diminished because "to the best of our knowledge, explosivity limits had not been reached" in any of the households where Range Resources briefly installed monitoring devices.

Range Resources Agreed to Conduct Further Research and Sampling, but Questions About Data Quality and Contamination Remain

Another reason for withdrawing the emergency order was that in 2012 the EPA had begun discussions with Range Resources that resulted in a non-binding agreement being reached in the spring of 2012. Under the agreement, Range Resources agreed to participate in the EPA's national study of drinking water effects from hydraulic fracturing once the EPA withdrew the order. Range Resources also agreed to test the water from 20 wells near its gas production well in Parker County for a year.

The EPA said that Range Resources' hydraulic fracturing facilities and operational information would make a valuable contribution to the study. Prior to reaching this agreement, Range Resources had indicated that they would not allow the EPA access to their facilities or participate in the study as long as the emergency order remained. However, as part of the agreement, Range Resources agreed to allow the EPA access to their gas facilities in Pennsylvania and would provide operational information to the agency. As of August 2013, the EPA and Range Resources had not agreed on the terms of Range Resources' participation. A senior EPA official said that the outcome of ongoing discussions between Range Resources and the EPA was uncertain.

Range Resources also agreed to test 20 wells near the gas well every 3 months for a year, and the EPA said that the 20 wells were close enough to the gas well that sample results would indicate whether there is a wider contamination problem. Sampling these wells would show whether there was an immediate risk to individual homeowners not included in the emergency order.

According to the EPA, the sampling that Range Resources has completed indicates no widespread methane contamination above action levels in the wells that were sampled in Parker County (only one well of 20 showed methane above that level, and a subsequent sample at this well was below that level). However, the agreement for testing did not include other elements of the original emergency order directed toward characterizing the ISE, such as testing the soil in the area of the contamination, conducting indoor air monitoring, conducting a geographical survey and defining contamination pathways. OECA managers accepted this partial solution because Range Resources would not voluntarily conduct these elements of the order, and they judged that the EPA could not spare the resources to continue enforcement through the courts.

In addition, the EPA lacks quality assurance information for Range Resources' sampling program, and questions remain about the presence of contamination. We identified two limitations with the approach taken. First, the sampling by Range Resources excluded one of the two wells where contamination was first identified in 2010. According to both the EPA and Range Resources, this was due to an ongoing lawsuit between the company and the homeowner. Second, Range Resources, via its attorney's March 30, 2012, letter, committed to sample the private water wells in accordance with EPA-sanctioned test methods. However, the EPA did not review or approve Range Resources' sampling protocol, nor did it review or approve the data collection and analytical methods during the course of the study.

In our draft report, we stated that the EPA should have an oversight role in Range Resources' sampling program. We believe this is important to ensure the validity and reliability of the data that the EPA is using to evaluate whether additional contamination of drinking water sources exists. Subsequent to the issuance of our draft report, EPA reviewed Range Resources' quarterly sampling data and determined that the data lacked some of the required quality assurance information. As a result, in an August 22, 2013, letter to Range Resources' attorney (see Appendix B), a Region 6 enforcement chief requested that Range Resources provide to Region 6 information on Range Resources' quarterly sampling, such as field results and sampling notes, sampling locations, sampling methods, and chain of custody records for sample results.

Conclusions

Region 6 and EPA headquarters met all requirements of Section 1431 of the SDWA in issuing the emergency order to Range Resources and in communicating with the public and stakeholders. In addition, since SDWA Section 1431 and the EPA's guidance do not provide criteria for withdrawing the emergency order, the EPA's use of discretion in withdrawing the order was allowed. The EPA withdrew the order because an agreement with Range Resources was underway; the costs and risks of litigating this particular case were likely to be very high and the needed short-term benefits would be low, if any; and immediate human health risks were believed to have been addressed.

One matter that led the EPA to issue the emergency order was addressed—Range Resources agreed to perform additional tests near the wells. However, other issues remain. Although EPA officials believe that current residents are not presently at risk, the overall risk faced by current and future area residents has not been determined. We believe that the EPA needs to implement cost-effective steps to better gauge the risk and document and disseminate its findings to affected residents. The EPA should oversee the sampling Range Resources agreed to perform and conduct its own sampling at any locations it suspects may be contaminated that are not contained in Range Resources' sample. The EPA should also provide homeowners with information about any contamination at their homes and work with the RRC to take corrective actions.

Recommendations

We recommend that the Regional Administrator, Region 6:

1. Collect and evaluate the testing results being provided by Range Resources to determine whether the data are of sufficient quality and utility.

2. Using quality data collected and analyzed, determine whether an ISE still exists at the original residential wells that prompted the emergency order and at other wells where Range Resources has identified contamination.

3. Advise residents at sites of evaluated wells of the present status of the contamination and of any Region 6 planned actions.

4. As needed, work with the RRC to ensure appropriate action is taken to respond to any ISE at the sites where contamination was identified. If the data and information available to Region 6 indicate no ISE, document and communicate that decision.

5. Document the costs and resources invested to complete the work included in these recommendations.

Agency Response and OIG Evaluation

We received a response to the draft report on September 3, 2013. In the written response, the agency agreed with and provided corrective actions that address recommendations 3, 4 and 5. These recommendations are resolved with corrective actions underway.

In a meeting to discuss the agency's comments on our report and response to the recommendations, we sought clarification on recommendations 1 and 2. Specifically, for recommendation 1, the OIG requested that the agency commit to evaluate the information that it receives, and take appropriate actions should it determine that the data are not sufficient for it to reach a conclusion concerning the level of contamination of the underground source of drinking water. EPA agreed that it will take appropriate steps should any of the information it receives indicate a potentially significant data quality concern.

For recommendation 2 the OIG requested that the agency commit to taking action should the data collected indicate ISE to other drinking wells in the involved area. EPA agreed that should any of the sampling data provided to the EPA by Range Resources reveal ISEs to other drinking wells in the involved area, the EPA will take appropriate action by the end of the first quarter of fiscal year 2014. Based on our discussions and agreements with the EPA, recommendations 1 and 2 are resolved and open with corrective actions underway.

Status of Recommendations and Potential Monetary Benefits

		RECOMMENDATIONS				POTENTIAL MONETARY BENEFITS (in $000s)	
Rec. No.	Page No.	Subject	Status[1]	Action Official	Planned Completion Date	Claimed Amount	Agreed-To Amount
1	19	Collect and evaluate the testing results being provided by Range Resources to determine whether the data are of sufficient quality and utility.	O	Regional Administrator, Region 6	12/31/13		
2	19	Using quality data collected and analyzed, determine whether an ISE still exists at the original residential wells that prompted the emergency order and at other wells where Range Resources has identified contamination.	O	Regional Administrator, Region 6	12/31/13		
3	19	Advise residents at sites of evaluated wells of the present status of the contamination and of any Region 6 planned actions.	O	Regional Administrator, Region 6	12/31/13		
4	19	As needed, work with the RRC to ensure appropriate action is taken to respond to any ISE at the sites where contamination was identified. If the data and information available to Region 6 indicate no ISE, document and communicate that decision.	O	Regional Administrator, Region 6	12/31/13		
5	19	Document the costs and resources invested to complete the work included in these recommendations.	O	Regional Administrator, Region 6	3/31/14		

[1] O = recommendation is open with agreed-to corrective actions pending
C = recommendation is closed with all agreed-to actions completed
U = recommendation is unresolved with resolution efforts in progress

Agency Response to Recommendations

UNITED STATES ENVIRONMENTAL PROTECTION AGENCY
1200 PENNSYLVANIA AVENUE, N.W.
WASHINGTON, D.C. 20460

September 3, 2013

MEMORANDUM

SUBJECT: Response to the Office of Inspector General Draft Report: "Response to Congressional Inquiry Regarding the EPA's Emergency Order to the Range Resources Gas Drilling Company," dated July 18, 2013, Project No. OPE-FY12-0019

FROM: Ron Curry
Regional Administrator
Region 6

Cynthia Giles
Assistant Administrator
Office of Enforcement and Compliance Assurance

TO: Carolyn Copper
Assistant Inspector General
Office of Program Evaluation

Thank you for the opportunity to respond to the draft findings and recommendations presented in the Office of Inspector General (OIG) Draft Report, "Response to Congressional Inquiry Regarding the EPA's Emergency Order to the Range Resources Gas Drilling Company." We appreciate the OIG's careful consideration of this matter and agree with your conclusion that EPA's issuance of the emergency order (order) was supported by law and fact, and that our exercise of discretion to resolve the matter was consistent with all applicable rules and policies.

In particular, the Draft Report finds that EPA satisfied all requirements of Section 1431 of the Safe Drinking Water Act, 42 U.S.C. § 300i, in issuing the order to Range Resources, and that issuance of the order was supported by the information in the Agency's possession, the location and timing of Range Resources' gas well drilling and gas production activities, and EPA's isotopic analysis. In addition, the Draft Report concludes that the order issued to Range Resources was similar to other recent Safe Drinking Water Act emergency orders, and that EPA's communications with the State and other stakeholders adhered to applicable law, policies, and procedures. The Draft Report also concludes that EPA consistently coordinated with the

Railroad Commission of Texas (Railroad Commission) prior to the order's issuance. Finally, the Draft Report finds that withdrawal of the order was within EPA's appropriate discretion and that the resolution of the matter did not violate any regulation or policy. We think that these conclusions are well supported and we agree.

While we agree with the overall conclusions of your review, there are a few minor places where the Draft Report is unclear or where the information differs from our understanding of the specific facts. We are attaching a list of these instances, along with suggested language to clarify or improve accuracy, for your consideration.

Responses to Recommendations

We appreciate the OIG's recommendations. Withdrawal of the order and resolution of the related federal litigation allowed the Agency to shift its focus in this particular matter towards a joint effort on the science and safety of energy extraction. Pursuant to its attorney's March 30, 2012 letter, Range Resources indicated its intent to conduct a year-long sampling effort (quarterly sampling events) in accordance with EPA-sanctioned test methods. We are committed to ensuring that all data and related information collected by Range Resources as part of its quarterly sampling events and provided to EPA are shared with the Railroad Commission, which is the lead state agency charged with overseeing oil- and gas-related activities in Texas. Furthermore, we will discuss the OIG's recommendations with the Railroad Commission and offer whatever assistance the Railroad Commission may require in carrying out its oversight functions in this area.

Recommendation 1: On August 22, 2013, we sent the attached letter to Range Resources' attorney regarding quality assurance information related to its quarterly sampling events. Upon further review of Range Resources' quarterly sampling data, we determined that the data contained some quality assurance information but lacked other such data. As described in our August 22 letter, we identified certain quality assurance information that appeared to be lacking and requested that Range Resources provide such additional information. The letter also requested that as part of its response, Range Resources confirm that it has submitted all information related to its quarterly sampling events. We will promptly share any additional information that we receive from Range Resources regarding its quarterly sampling events with the Railroad Commission.

Recommendation 2: Since EPA took samples in 2010 that ultimately led to the issuance of the order, circumstances have changed. Based on information available to the agency, the one private water well that was of primary concern has been disconnected from the home thereby addressing explosivity concerns, and the well is no longer used as a source of drinking water. In addition, only one of the approximately 80 water well samples taken by Range Resources identified methane above the action level of 10 milligrams per liter (mg/L), which is the level at which the U.S. Geological Survey (USGS) recommends having water wells evaluated for possible venting (the subsequent quarterly sample taken from this well identified methane below 10 mg/L). Importantly, EPA retains authority to take action in the future should circumstances change and the Railroad Commission fails to act as the lead state agency charged with overseeing oil- and gas-related activities in Texas.

Furthermore, we respectfully suggest that this recommendation is beyond the scope and purpose of this investigation, which was described on page five of the Draft Report.

Recommendation 3: We have taken appropriate action with respect to Range Resources' analyzed quarterly sampling data by sharing it with the Railroad Commission, which is the lead state agency charged with overseeing oil- and gas-related activities in Texas. Our understanding is that it is the Railroad Commission's practice to notify water well owners when the State possesses data that suggests contamination levels of potential concern. As noted above, we will promptly share any additional information that we receive from Range Resources regarding its quarterly sampling events with the Railroad Commission.

Recommendation 4: We have taken appropriate action with respect to Range Resources' analyzed quarterly sampling data by sharing it with the Railroad Commission. The analyzed quarterly sampling data submitted by Range Resources indicate that there is not widespread groundwater contamination of concern in the wells that were sampled in Parker County. As discussed above, only one of the approximately 80 water well samples taken by Range Resources identified methane levels above the 10 mg/L level identified by USGS, with the subsequent quarterly sample from that well identifying methane below 10 mg/L. We have shared these data with the Railroad Commission for them to follow up to ensure that any elevated methane levels do not pose a concern. We will promptly share any additional information that we receive from Range Resources regarding its quarterly sampling events with the Railroad Commission.

Recommendation 5: We will document the costs associated with reviewing any additional information that we receive from Range Resources regarding its quarterly sampling events and coordinating with the Railroad Commission.

Responses to Recommendations Table

No.	Recommendation	High-Level Intended Corrective Action(s)	Estimated Completion by Quarter and FY
1	Collect and evaluate the testing results being provided by Range Resources to determine whether the data is of sufficient quality and utility.	We have re-evaluated quality assurance information associated with Range Resources' analyzed quarterly sampling data and sent the attached letter to Range Resources' attorney. We will promptly share any additional information that we receive from Range Resources with the Railroad Commission.	First Quarter FY 2014 upon receipt of any additional information.
2	Using quality data collected and analyzed, determine whether an ISE still exists at the original residential wells that prompted the emergency order and at other wells where Range Resources has identified contamination.	As noted above, the well of primary concern has been disconnected from the home thereby addressing explosivity concerns, and it is no longer a source of drinking water. EPA retains the authority to take action if circumstances change in Parker County and the Railroad Commission fails to act as the lead state agency charged with overseeing oil- and gas-related activities in Texas.	No further action proposed.
3	Advise residents at sites of evaluated wells of the present status of the contamination and of any Region 6 planned actions.	We have shared all of Range Resources' analyzed quarterly sampling data with the Railroad Commission. We will promptly share any additional information that we receive from Range Resources regarding its quarterly sampling events with the Railroad Commission.	First Quarter FY 2014 upon receipt of any additional information.

No.	Recommendation	High-Level Intended Corrective Action(s)	Estimated Completion by Quarter and FY
4	As needed, work with the RRC to ensure appropriate action is taken to respond to any ISE at the sites where contamination was identified. If the data and information available to Region 6 indicate no ISE, document and communicate that decision.	We have shared all of Range Resources' analyzed quarterly sampling data with the Railroad Commission. The analyzed quarterly sampling data indicate that there is not widespread groundwater contamination of concern in the wells that were sampled in Parker County. We will promptly share any additional information that we receive from Range Resources with the Railroad Commission.	First Quarter FY 2014 upon receipt of any additional information.
5	Document the costs and resources invested to complete the work included in these recommendations.	We will document any costs and resources expended to address these recommendations.	Second Quarter FY 2014.

Contact Information

If you have any questions or concerns regarding this response, please contact the Region 6 Audit Liaison, Susan Jenkins, at (214) 665-6578.

Attachments

cc: Charles Sheehan, OIG
        Samuel Coleman, R6
        John Blevins, R6/CAED
        Stephen Gilrein, R6/CAED
        Jerry Saunders, R6/CAED
        Steven Chester, OECA
        Susan Shinkman, OECA/OCE
        Pamela Mazakas, OECA/OCE
        Andrew Stewart, OECA/OCE
        Timothy Sullivan, OECA/OCE
        Lauren Kabler, OECA/OCE
        Gwendolyn Spriggs, OECA/OAP

Agency Letter to Range Resources

UNITED STATES ENVIRONMENTAL PROTECTION AGENCY
REGION 6
1445 ROSS AVENUE, SUITE 1200
DALLAS, TEXAS 75202 – 2733

August 22, 2013

Mr. John Riley
Bracewell & Giuliani
111 Congress Avenue
Suite 2300
Austin, TX 78701-4061

Dear Mr. Riley:

Thank you for undertaking the four quarterly sampling events and for providing information in a timely manner related to those events as committed to in your letter dated March 30, 2012, to Mr. Steven Chester regarding Range Production Company and Range Resources Corporation (Range). It is my understanding that all of the sampling events have been conducted, and all analyses are complete.

The EPA is reviewing the information Range has submitted related to the sampling events (as discussed in paragraph 2 and 3 of your letter) that will help the EPA and the State of Texas (The Railroad Commission of Texas) assess the quality of the data collected and the integrity of the sampling. Based upon a review of the data provided by Range, it does not appear that we have the following information regarding the quarterly sampling:

a. Field results and records made during sampling. (We have a copy of field results for the first quarter sample event, but we do not have the field records and results for the last three quarters.);

b. Trip blanks, equipment blanks, duplicates, matrix spikes and matrix spike duplicates. (These were taken, but the results were not submitted.);

c. Chain of custody forms for samples sent to Isotech. (These were only provided for the first quarter. We have not received these forms for quarters 2, 3 and 4.);

d. Sample locations were not provided for the 4th quarter sampling event, and a summary table was not provided.); and

e. Description of the sampling technique for dissolved gas, including the amount of time between filling the sample bottles, wetting the cap and placing the cap on the bottle.

Additionally, if Range has any further information related to the quarterly sampling events and analyses pursuant to the March 30, 2012 letter, please submit this as well. All information should be sent to Jerry Saunders in the Region 6 office within 30 days of the receipt of this letter. The Region has shared all of the information submitted to date by Range with the Railroad Commission of Texas, the primary regulatory agency for the wellfield in southern Parker County. We will share any additional data submitted as well.

2

As part of your response, please confirm in your reply that Range has submitted the entirety of this information. If you have any questions regarding this matter, do not hesitate to call me at 214-665-2718. Thank you for your cooperation.

Sincerely,

Scott McDonald, Chief
Water Enforcement Branch
Office of Regional Counsel

cc: Mr. David Poole
Range Production Company

Mr. Peter Pope
Railroad Commission of Texas

Mr. Steven Chester
Office of Enforcement and Compliance Assurance

OIG Response to Agency Comments

OIG Recommendation	EPA Response	EPA Corrective Action	OIG Assessment of EPA Response
1. Collect and evaluate the testing results being provided by Range Resources to determine whether the data are of sufficient quality and utility.	"We have re-evaluated quality assurance information associated with Range Resources' analyzed quarterly sampling data and sent the attached letter to Range Resources' attorney. We will promptly share any additional information that we receive from Range Resources with the Railroad Commission." NOTE: In a subsequent communication, EPA agreed to take appropriate steps should any of the information it receives indicate a potentially significant data quality concern.	First Quarter FY 2014 upon receipt of any additional information.	We agree with the EPA's proposed actions. This recommendation is resolved.
2. Using quality data collected and analyzed, determine whether an ISE still exists at the original residential wells that prompted the emergency order and at other wells where Range Resources has identified contamination.	"As noted above, the well of primary concern has been disconnected from the home thereby addressing explosivity concerns, and it is no longer a source of drinking water. EPA retains the authority to take action if circumstances change in Parker County and the Railroad Commission fails to act as the lead state agency charged with overseeing oil- and gas-related activities in Texas." NOTE: In a subsequent communication, EPA agreed to take appropriate steps should any of the sampling data collected by Range Resources reveal imminent and substantial risks to other drinking wells in the involved area.	First Quarter FY 2014.	The OIG accepts the agency's interpretation that the well of primary concern has been disconnected from the home, thereby addressing explosivity concerns, and it is no longer a source of drinking water. We agree with the EPA's proposed actions. This recommendation is resolved.

OIG Recommendation	EPA Response	EPA Corrective Action	OIG Assessment of EPA Response
3. Advise residents at sites of evaluated wells of the present status of the contamination and of any Region 6 planned actions.	"We have shared all of Range Resources' analyzed quarterly sampling data with the Railroad Commission. We will promptly share any additional information that we receive from Range Resources regarding its quarterly sampling events with the Railroad Commission."	First Quarter FY 20014 upon receipt of any additional information.	We agree with the EPA's proposed actions. This recommendation is resolved.
4. As needed, work with the RRC to ensure appropriate action is taken to respond to any ISE at the sites where contamination was identified. If the data and information available to Region 6 indicate no ISE, document and communicate that decision.	"We have shared all of Range Resources' analyzed quarterly sampling data with the Railroad Commission. The analyzed quarterly sampling data indicate that there is not widespread groundwater contamination of concern in the wells that were sampled in Parker County. We will promptly share any additional information that we receive from Range Resources with the Railroad Commission."	First Quarter FY 2014 upon receipt of any additional information.	We agree with the EPA's proposed actions. This recommendation is resolved.
5. Document the costs and resources invested to complete the work included in these recommendations.	"We will document any costs and resources expended to address these recommendations."	Second Quarter FY 2014.	We agree with the EPA's proposed actions. This recommendation is resolved.

Distribution

Office of the Administrator
Assistant Administrator for Enforcement and Compliance Assurance
Regional Administrator, Region 6
Agency Follow-Up Official (the CFO)
Agency Follow-Up Coordinator
General Counsel
Associate Administrator for Congressional and Intergovernmental Relations
Associate Administrator for External Affairs and Environmental Education
Principal Deputy Assistant Administrator for Enforcement and Compliance Assurance
Deputy Regional Administrator, Region 6
Director, Office of Regional Operations
Region 6 Public Affairs Office
Region 6 Director for Enforcement
Audit Follow-Up Coordinator, Office of Enforcement and Compliance Assurance
Audit Follow-Up Coordinator, Region 6

www.ingramcontent.com/pod-product-compliance
Lightning Source LLC
Chambersburg PA
CBHW081242170526
45165CB00009B/3163